Loss of Start-Up Oxygen in CSE SR-100 Self-Contained Self-Rescuers

Department of Health and Human Services
Centers for Disease Control and Prevention
National Institute for Occupational Safety and Health

Loss of Start-Up Oxygen in CSE SR-100 Self-Contained Self-Rescuers

Robert Stein, Heinz Ahlers, Roland Berry Ann

DEPARTMENT OF HEALTH AND HUMAN SERVICES
Centers for Disease Control and Prevention
National Institute for Occupational Safety and Health

April 2012

This document is in the public domain and may be freely copied or reprinted.

Disclaimer

Mention of any company or product does not constitute endorsement by the National Institute for Occupational Safety and Health (NIOSH). In addition, citations to Web sites external to NIOSH do not constitute NIOSH endorsement of the sponsoring organizations or their programs or products. Furthermore, NIOSH is not responsible for the content of these Web sites. All Web addresses referenced in this document were accessible as of the publication date.

Ordering Information

To receive documents or other information about occupational safety and health topics, contact NIOSH at

> Telephone: **1–800–CDC–INFO** (1–800–232–4636)
> TTY: 1–888–232–6348
> e-mail: cdcinfo@cdc.gov
>
> or visit the NIOSH Web site at **www.cdc.gov/niosh**.

For a monthly update on news at NIOSH, subscribe to NIOSH *eNews* by visiting **www.cdc.gov/niosh/eNews**.

DHHS (NIOSH) Publication No. 2012–139

April 2012

SAFER • HEALTHIER • PEOPLE™

Loss of Start-Up Oxygen in CSE SR-100 Self-Contained Self-Rescuers

Robert Stein, Heinz Ahlers, Roland Berry Ann

Executive Summary

This report describes a National Institute for Occupational Safety and Health (NIOSH) and Mine Safety and Health Administration (MSHA) investigation assessing the prevalence of a lack of sufficient start-up oxygen in CSE SR-100 self-contained self-rescuer (SCSR) devices.

The availability of sufficient start-up oxygen is critical to the performance of the SR-100. As part of a routine field testing program of SCSRs used in coal mines, NIOSH and MSHA detected two SR-100s that lacked sufficient start-up oxygen. CSE Corporation subsequently discovered one SCSR that lacked sufficient start-up oxygen in that company's internal quality control program and voluntarily stopped further production and sales of SR-100s.

NIOSH developed a protocol to test for the presence of start-up oxygen in field-deployed SR-100s. The purpose of the test was to determine if the failure rate of the start-up oxygen in the population of 70,000 field-deployed units exceeded 1%. NIOSH and MSHA used American Society for Quality (ASQ), Sampling Procedures and Tables for Inspection of Isolated Lots by Attributes (ASQC Q3-1988)[1]. In assessing the SR-100s, if no more than 3 failures of start-up oxygen occurred in the 500-unit random sample, the SR-100 could be accepted as meeting the Limiting Quality (LQ) rate of 1.25% for start-up oxygen performance.

NIOSH tested five hundred field-deployed devices collected from coal mines throughout the United States. NIOSH observed 5 start-up oxygen failures in the 500 units it tested. The maximum number of failures allowed under the LQ rate of 1.25% was exceeded; therefore, the 1% maximum allowable failure rate under the protocol was not met.

---

[1] American National Standard, Sampling Procedures and Tables for Inspection of Isolated Lots by Attributes, (ANSI/ASQC Standard Q3-1988), American Society for Quality, 1988.

Contents

Executive Summary iii
Contents iv
Acronyms and Abbreviations vi
Introduction 1
Methods 2
Results 4
Major Findings 9
Appendix 1 10

Figures

Figure 1. Age histogram of SR-100 population 5
Figure 2. Age histogram of targeted sample 5
Figure 3. Age histogram of tested sample 6
Figure 4. Unit 154821 8
Figure 5. Unit 154821 – Dented bottom cover. 8
Figure 6. Unit 249530 8
Figure 7. Unit 114752 8
Figure 8. Unit 251475 9
Figure 9. Unit 249607 9

Tables

Table 1. Inspection Failures 6
Table 2. Start-up oxygen failure details 7

Acronyms and Abbreviations

ANSI	American National Standards Institute
AQL	Acceptable Quality Level
ASMD	Acoustic Solids Movement Detector
ASQ	American Society for Quality
ASQC	American Society for Quality Control
CFR	Code of Federal Regulations
LQ	Limiting Quality
LTFE	Long-Term Field Evaluation
MSHA	Mine Safety and Health Administration
NIOSH	National Institute for Occupational Safety and Health
QA	Quality Assurance
SCSR	Self-Contained Self-Rescuer
S/N	Serial Number
°F	Degrees Fahrenheit

Introduction

The National Institute for Occupational Safety and Health (NIOSH) and the Mine Safety and Health Administration (MSHA) jointly approve respirators for use in the nation's mines. Under 42 CFR[2] § 84.3 ("Respirators for mine rescue or other emergency use in mines"), NIOSH and MSHA must jointly review and issue certificates of approval for respirators used for mine emergencies and mine rescue, which includes the CSE SR-100, a one-hour SCSR. At the time of this investigation, the SR-100 was one of three one-hour SCSR models used in the nation's mines.

The SR-100 was approved in 1989 and has since been deployed in underground coal and metal/nonmetal mines for use in emergency escapes. CSE classified the operation of the start-up oxygen as a *Critical* attribute. The regulation (in section 84.41) classifies potential defects as follows:

> (1) *Critical.* A defect that judgment and experience indicate is likely to result in a condition immediately hazardous to life or health for individuals using or depending upon the respirator;
> (2) *Major A.* A defect, other than critical, that is likely to result in failure to the degree that the respirator does not provide any respiratory protection, or a defect that reduces protection and is not detectable by the user;
> (3) *Major B.* A defect, other than Major A or critical, that is likely to result in reduced respiratory protection, and is detectable by the user; and
> (4) *Minor.* A defect that is not likely to materially reduce the usability of the respirator for its intended purpose, or a defect that is a departure from established standards and has little bearing on the effective use or operation of the respirator.

To ensure the ongoing reliability of approved mine escape respirators, NIOSH and MSHA conduct a field testing program, referred to as the Long-Term Field Evaluation (LTFE), of SCSRs used in coal mines. Since 1982, the LTFE has identified reliability issues among the approved respirators stemming either from the environment or manufacturing. NIOSH notified CSE in December 2009 that two SR-100s exhibited little or no start-up oxygen during performance testing in the most recent phase of the LTFE. CSE's ensuing investigation identified excessive heat as a contributor to loss of start-up oxygen. However, the heat indicators on the failed units did not indicate prior exposure to excessive heat, suggesting a factor other than excessive heat was responsible for the loss of start-up oxygen.

---

[2] CFR. Code of Federal Regulations. Washington, DC: U.S. Government Printing Office, Office of the Federal Register.

During the ongoing investigation, CSE reported to NIOSH and MSHA that routine quality-assurance testing found an oxygen cylinder in a newly assembled SR-100 that contained an insufficient quantity of start-up oxygen. CSE immediately and voluntarily ceased production and sales of the SR-100. CSE subsequently reinspected a large quantity of the pre-filled oxygen cylinders not yet installed in SR-100s. These oxygen cylinders had been previously confirmed to have oxygen volume sufficient for the start-up function. But upon reinspection, CSE found between 0.5% and 1% of these pre-filled cylinder assemblies failed inspection due to oxygen volume loss.

CSE engaged an engineering firm to analyze the loss of start-up oxygen. The engineering firm concluded that the loss of oxygen occurred at a threaded connection between the cylinder body and the stopper/outlet assembly. When asked to determine the percentage of affected, assembled SR-100s, CSE responded that it was less than 1% overall. CSE could not identify a systemic cause or otherwise confine the failure within certain lots. Therefore, the failure could exist among all field-deployed units.

NIOSH developed a peer-reviewed protocol to evaluate a representative sample of field-deployed SR-100 respirators to determine the extent of the non-conformance. NIOSH and MSHA made a determination for coal-mine deployment that the failure of the start-up oxygen cylinder should be 1% or less in SR-100s deployed in U.S. coal mines. This criterion was based on MSHA's rules for coal mining requiring that each miner have an immediately available back-up unit.

NIOSH and MSHA collected more than 500 SR-100s from coal mines throughout the country. NIOSH tested for low start-up oxygen cylinders from 500 units meeting the inspection criteria. The purpose of these tests was to determine if the failure rate of the start-up oxygen exceeded 1% in the population of 70,000 field-deployed units. The remainder of this report describes the methods, results, major findings and the test protocol used for the investigation.

Methods

Sample Size Determination

Information collected and provided by MSHA in its SCSR inventory[3] indicated approximately 70,000 SR-100s were field-deployed during the time of this evaluation. A study protocol was developed that included a sampling strategy and a testing protocol (see Appendix I). The sampling plan was designed to detect with 95% confidence that the prevalence of cylinders with low start-up oxygen (i.e., failure rate) was less than 1.25% for a population of 70,000 units. The sample selection plan was based on American National Standards (ANSI)/American Society of

---

[3] Further information may be found at http://www.msha.gov/forms/ELawsForms/2000-222.htm

Quality (ASQC) Q3-1988, Sampling Procedures and Tables for Inspection of Isolated Lots by Attributes (Q3-1988), a quality assurance technique of sampling by attributes. The Limiting Quality (LQ) point nearest to and including the 1% criterion was selected from established tables with a sample size of 500 units. Three or fewer defective units would mean the failure rate for the population of 70,000 was less than or equal to 1.25% with 95% confidence.

Recoverable Unit Criteria

Each unit was assessed for compliance with manufacturer inspection criteria. The inspected characteristics assessed were the Acoustic Solids Movement Detector level (ASMD), moisture indicators, heat indicator, and several specific physical damage characteristics, which are grouped together and referred to as "damage."

The target sample size was set at 500 recoverable units (see Appendix 1). A recoverable unit was defined as a unit that could be located, would be relinquished by the owner, and passed all manufacturer inspection criteria for use. Prior experience collecting respirators from mine service indicated that more than 500 units would need to be collected and assessed to achieve the desired sample size. To adjust to difficulty encountered in collecting SR-100s, units were collected and tested in two major phases. However, the testing procedure did not change from phase 1 to phase 2.

Phase 1 Unit Collection and Testing

Units were randomly selected by serial number from the full list of 70,000 deployed units consistent with the original protocol (see Appendix 1). Compensation was offered to mine operators for any collected unit. Compliance with NIOSH/MSHA approval requires that owners/users of approved respirators abide by all manufacturer recommendations. Units passing all visual inspection and test criteria were considered recoverable and tested according to the test plan analysis described in the protocol (Appendix 1). Phase 1 collection and testing of SR-100s occurred over a period of 5 months.

Phase 2 Unit Collection and Testing

MSHA regulations established at 30 CFR § 75.1714-4, promulgated under the Mine Improvement and New Emergency Response Act of 2006, Pub. L. 109-236 (S. 2803), Section 2 (3)(C)(iii)(II), require coal mine operators to maintain specific quantities (multiple units per miner, per each operator's approved Emergency Response Plan) of SCSRs to support mine escape. Operators were concerned that providing units for testing would result in noncompliance with MSHA requirements for SCSR coverage before replacement units became available, and were thus unwilling to provide units for NIOSH testing. As such, it was not possible to complete the collection and testing of SR-100s within a reasonable period of time, using a pure random collection method.

NIOSH subsequently met with the original protocol peer reviewers to discuss the collection challenges and determine an alternate strategy. The reviewers agreed that little if any bias would be introduced if "availability" was the priority for collecting the remaining units. Availability, as used here, indicates the units could be located and their owners were able to part with them through the NIOSH replacement offers. This protocol change allowed NIOSH to collect recoverable units from coal mine operators who were able to part with their units and still meet MSHA backup unit requirements.

NIOSH and MSHA identified two large mine operators who were in the process of withdrawing SR-100s on a mine-by-mine basis. Units were selected by obtaining the highest ranking serial numbered units in the randomized list of 70,000 deployed units. This resulted in sporadic concentrations of production months, but the overall sample still spans the time period over which CSE produced SR-100s under the prevailing construction method. Similar to phase 1, collected units were inspected according to the manufacturer criteria. Phase 2 collection and testing of SR-100s occurred over a period of 3 months.

Test Method

Each recoverable unit was catalogued and photographed. The units were prepared for oxygen volume evaluation. The start-up oxygen volume was released, measured and compared to the manufacturer's established QA criteria. Failure was determined when the volume of start-up oxygen released was less than the age-adjusted minimum volume defined in the CSE quality assurance plan (Appendix 1). This difference is noted as oxygen volume loss in Table 2.

Results

Age Distribution of Units

Figures 1, 2, and 3 illustrate the manufacturing date frequency among the population at large (Figure 1), the targeted population (Figure 2), and the tested sample (Figure 3).

A total of 137 SR-100s were collected in phase 1. One hundred and nine units were determined to be recoverable and subsequently tested, and 28 units were rejected due to one or more failures during inspection (see Table 1). No units were presented with failing moisture indicators, and few were presented with unacceptable amounts of physical damage.

During the second phase of collection, units failing the manufacturer's inspection criteria were rejected from inclusion in the oxygen starter study, but the reasons for rejection were no longer analyzed or tallied. Three hundred ninety-one recoverable units were identified and tested according to the test plan.

The study was not designed to evaluate the distribution and rate at which field-deployed units failed to meet the manufacturer's standards for continued deployment.

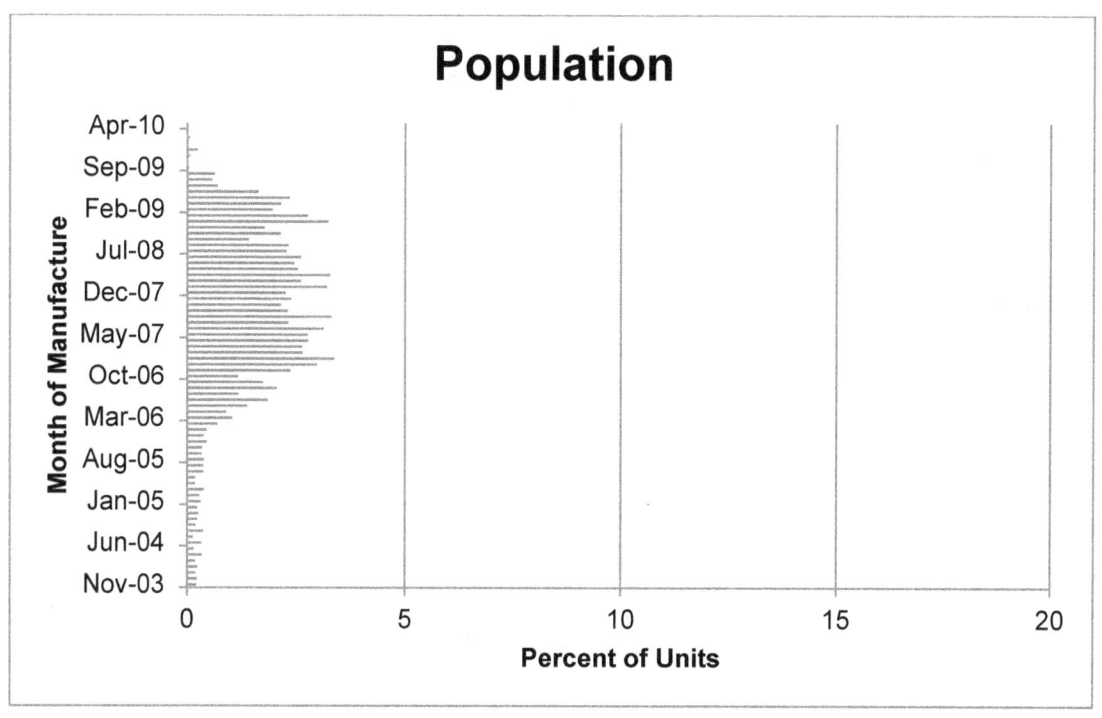

Figure 1. Age histogram of SR-100 population

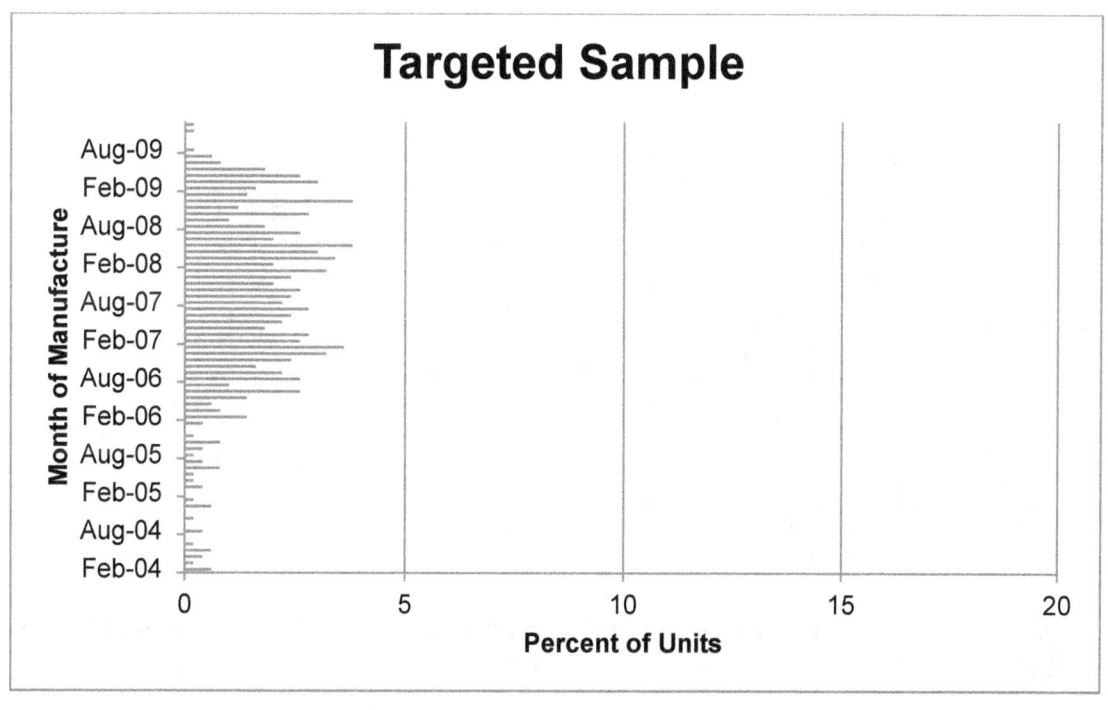

Figure 2. Age histogram of targeted sample

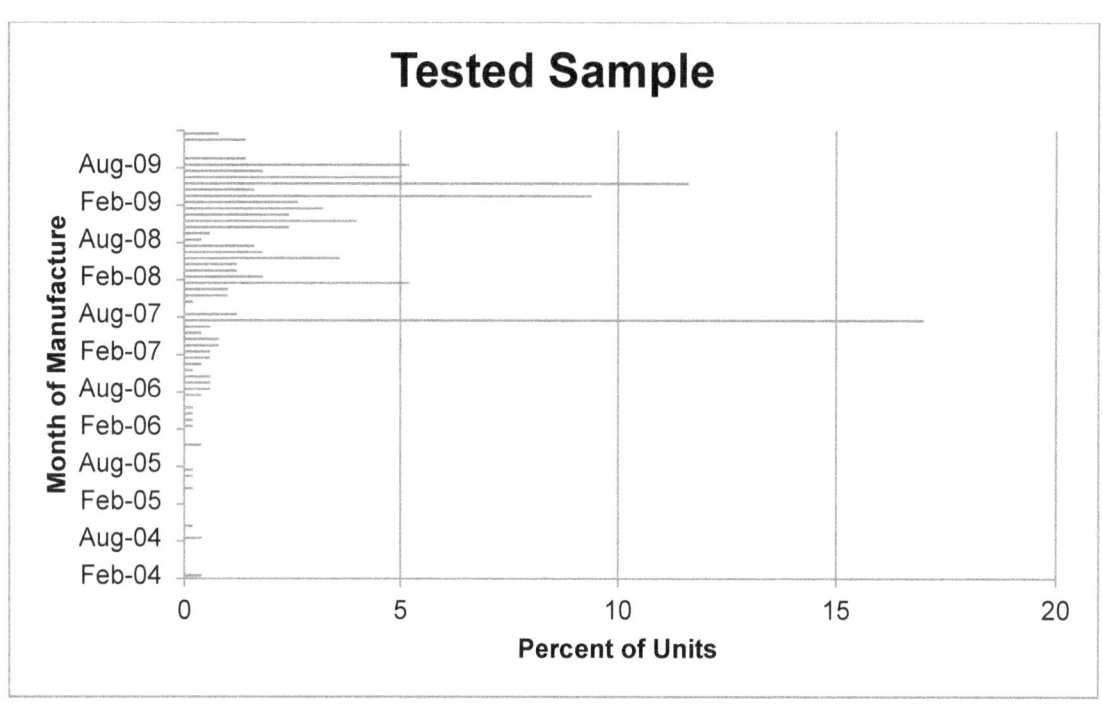

Figure 3. Age histogram of tested sample

Table 1. Inspection Failures

Serial No.	Total	104922	110284	111535	113083	116528	117030	120205	124293	130302	137049	137099	137613	145254	146851	168662	181851	187725	187902	189276	191208	202636	148400	148445	206317	209424	223677	231516	231542
ASMD	20	Fail	Fail	Fail		Fail	Fail	Fail	Fail	Fail	Fail		Fail	Fail	Fail	Fail	Fail	Fail	Fail			Fail	Fail		Fail	Fail	Fail	Fail		
Heat Indicator	14		Fail	Fail			Fail	Fail	Fail	Fail	Fail								Fail	Fail	Fail		Fail			Fail	Fail	Fail	Fail	
Damage	4													Fail												Fail	Fail			Fail

ASMD = Acoustic Solids Movement Detector
137 units were inspected during phase 1

All recoverable units were tested according to the procedures outlined in the protocol (Appendix 1). NIOSH observed five (5) failures among the 500 units tested (see Table 2).

Table 2. Start-up oxygen failure details

Test Date	Unit Serial Number	Cylinder Serial Number	Manufacture Date	Volume of Oxygen (liters)	QA Minimum Volume Oxygen (liters)	Oxygen Volume Deficiency (liters)	Pass/Fail	Age at Test Date (Months)	Stored/Carried (Observation of current condition)
02-08-11	154821	A68341	02-2007	3.97	6.4	2.43	Fail	48	Carried
04-19-11	249530	A181402	06-2009	5.20	7.4	2.20	Fail	22	Carried
04-19-11	114752	A20125	04-2005	0.02	6	5.98	Fail	72	Carried
04-29-11	251475	A180503	08-2009	6.09	7.5	1.41	Fail	20	Carried
05-17-11	249607	A181317	06-2009	0.93	7.4	6.47	Fail	23	Carried

Overall, the failure discovery rate was 1 per 100 units examined. This rate was observed in both phases of collection, also suggesting the change in collection methodology did not affect the evaluation. One unit failed among the 109 tested from collection in phase 1, and 4 units failed among the 391 tested from collection in phase 2.

The 500 tested SR-100s may be viewed as representing three sets of deployed units:
  (1) SR-100 units deployed at all coal mines using SR-100s (109 collected units, 1 failure detected),
  (2) SR-100 units deployed at coal mines operated by two targeted mining companies (391 collected units, 4 failures detected), and
  (3) SR-100 units from the combination of sets 1 and 2 (500 collected units, 5 failures detected).

The combined set (3) is considered to be representative of the population of all deployed units.

Recoverable units were photographed prior to being opened for testing. Figures 4 through 9 are photographs of the units with insufficient start-up oxygen. The photographs illustrate typical conditions of conforming fielded units. Many other tested units with sufficient start-up oxygen appeared to have incurred more wear and tear than the average failed unit.

Figure 4. Unit 154821

Figure 5. Unit 154821 – Dented bottom cover.

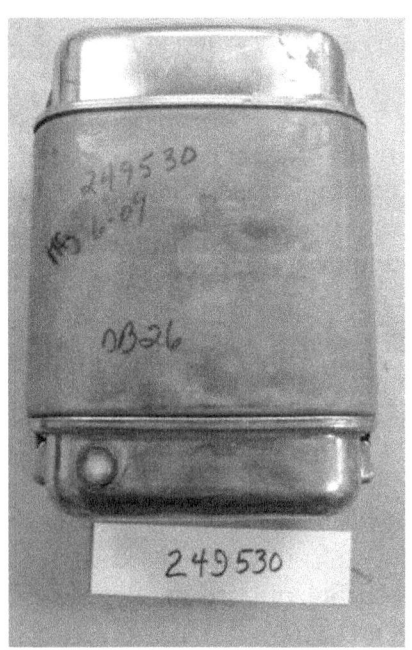

Figure 6. Unit 249530

Figure 7. Unit 114752

Figure 9. Unit 251475 **Figure 8.** Unit 249607

Major Findings

All of the units in which excessive start-up oxygen loss was observed had been carried for at least some portion of their deployed lives, and exhibited typical conditions of conforming fielded units. While the unit with the largest total start-up oxygen loss was the oldest of this group, the next largest total loss measured was in one of the two newest units.

NIOSH observed 5 start-up oxygen failures in the 500 units it tested. By the criteria set forth in the ASQC sampling tables, the target LQ of 1.25% is not met. The allowable failure rate of 1% established under the protocol is exceeded.

Appendix 1

Protocol for Sampling, Testing and Analyzing Oxygen-Starter Performance of the CSE SR-100 Self-Contained Self-Rescuer

Purpose:

The purpose of this protocol is to establish procedures to sample, test, and analyze the CSE SR-100 Self-Contained Self-Rescuer (SCSR) oxygen starter performance in mine-deployed respirators.

Background:

The primary oxygen supply of the CSE, SR-100 SCSR is stored chemically. Upon activation, the SR-100 oxygen starter is designed to provide approximately 9 liters of gaseous oxygen into the breathing circuit of the SCSR. The injection of gaseous starter oxygen provides initial oxygen for breathing to allow time for the reaction in the chemical bed to activate and begin producing oxygen for the user. Tests performed as part of the National Institute for Occupational Safety and Health (NIOSH) Long Term Field Evaluation audit program in December 2009 and by the manufacturer as part of its quality assurance program, in February 2010, revealed failures of oxygen starters on some units. CSE has calculated that the oxygen-starter failure rate on new units is less than one percent. CSE further reported to NIOSH and the Mine Safety and Health Administration (MSHA) that the oxygen starter failure is related to application of thread sealant and thread dimensions at the starter oxygen high-pressure valve connection and that the valve connection assembly process is not traceable to a manufacturing lot or manufacturing date. The actual oxygen-starter failure rate in field-deployed units is not well characterized at this time. Based on certification performance requirements established at 42 CFR pt. 84, NIOSH and MSHA have determined that only a fully functioning oxygen starter enables performance of the SR-100 to be operationally compliant. Starter oxygen is required for proper function in the first several minutes of use and to ensure a complete oxygen supply. The 42 CFR pt. 84, sampling plans for addressing quality characteristics and the associated manufacturing process accepted quality levels (AQLs) are found at 42 CFR § 84.41:

> § 84.41 Quality control plans; contents.
> (c) The sampling procedure shall include a list of the characteristics to be tested by the applicant or his agent.
> (d) The characteristics listed in accordance with paragraph (c) of this section shall be classified according to the potential effect of such defect and grouped into the following classes:
> (1) *Critical.* A defect that judgment and experience indicate is likely to result in a condition immediately hazardous to life or health for individuals using or depending upon the respirator;
> (2) *Major A.* A defect, other than critical, that is likely to result in failure to the degree that the respirator does not provide any respiratory protection, or a defect that reduces protection and is not detectable by the user;

(3) *Major B.* A defect, other than Major A or critical, that is likely to result in reduced respiratory protection, and is detectable by the user; and

(4) *Minor.* A defect that is not likely to materially reduce the usability of the respirator for its intended purpose, or a defect that is a departure from established standards and has little bearing on the effective use or operation of the respirator.

(e) The quality control inspection test method to be used by the applicant or his agent for each characteristic required to be tested shall be described in detail.

(f) Each item manufactured shall be 100 percent inspected for defects in all critical characteristics and all defective items shall be rejected.

(g) The Acceptable Quality Level (AQL) for each major or minor defect so classified by the applicant shall be:

(1) *Major A.* 1.0 percent;

(2) *Major B.* 2.5 percent; and

(3) *Minor.* 4.0 percent.

According to the NIOSH/MSHA approved quality plan, CSE defines the supply of starter oxygen for the SR-100 as a critical attribute. The expected failure rate of critical attributes is zero. Test data provided by CSE indicate the actual failure rate to be around one percent. Redundancy in deployment of SCSRs required under 30 CFR § 75.1714-4, promulgated under the Mine Improvement and New Emergency Response Act of 2006, Pub. L. 109-236 (S. 2803), Section 2 (3)(C)(iii)(II), helps to offset any negative impact from unexpected performance problems and in this situation is being relied upon as a temporary solution. In a user notice offered by CSE on May 10, 2010, CSE acknowledged the benefit of redundancy in the event of oxygen starter failure, advising that, "If for any reason a unit does not inflate the breathing bag, the user should don another unit if one is readily available. If a second unit is not readily available, the manual start should be used." If an additional SCSR is readily available, the existing data showing a 1 in 100 (1%) failure rate would yield a 1 in 10,000 chance of having both units fail in any sequence of two SCSRs.

As a permanent solution, CSE has proposed redesign of the SR-100, but as of the date of this protocol has not submitted plans to NIOSH and MSHA for approval test and evaluation. However, none of the improvements proposed thus far will address units currently deployed. Both NIOSH and MSHA expect that currently deployed SR-100s will ultimately need to be replaced, and that the pace of the replacement needs to be driven by the actual oxygen-starter failure rate. To that end, a sampling of field-deployed SR-100s is proposed to determine the prevalence of failed oxygen starters in the CSE SR-100 respirators.

Approach

In order to properly establish the actual prevalence of failed oxygen starters among field-deployed SR-100s, NIOSH proposes using a quality assurance (QA) approach. Operation of the oxygen starter is a single quality attribute. Defined in this manner, standard QA sampling methods may be relied upon to yield a statistically significant characterization of the overall number of failures by the number of failures observed in a reasonably small sample. The total mine-deployed population of SR-100s currently exceeds 70,000 units, thus it is crucially important to draw upon widely recognized sampling techniques and statistical methods. While the method proposed is not designed to find the actual proportion of failed starters, it will with great certainty establish important limits. For example, it will not be possible to state the failure rate is 0.8%, but it will be possible to state the failure rate does not exceed 1.25%, at a 95% confidence interval. The criteria selected are based around limits of 1.25%, 5%, 8%, and 12.5%.

Criteria

The SR-100 sample selection plan is based on American National Standards (ANSI)/American Society of Quality (ASQC) Q3-1988, Sampling Procedures and Tables for Inspection of Isolated Lots by Attributes (Q3-1988). NIOSH regulation 42 CFR § 84.43 lists Mil-Std-105D as an acceptable sampling plan. Mil-STD-105D, as well as the now more widely used ANSI Z1.4, provides for sampling by Limiting Quality (LQ) levels when the actual consumer risk is at issue. The Q3-1988 is a recognized standard for LQ sampling plans. Q3-1988 plan offers LQ rates for 0.5%, 0.8% and 1.25% nominal limiting quality. It also offers higher LQ rates up to 32%. An LQ rate of 1.25% represents a 95% confidence level that the lot will have less than 1.25% for the tested characteristic. Thus, a LQ value of 1.25% is a reasonable approximation of a 1% consumer risk defect level.

To evaluate deployed SR-100 units for sufficient starter oxygen, NIOSH proposes to use Q3-1988 Limiting Quality (LQ) values. As noted, Q3-1988 lists standard tables of LQ quality levels ranging from 0.05 to 32 percent for various lot sizes. For a quantity of 35,000 to 100,000 deployed SR-100 respirators and LQ's ranging from 1.25 to 12.5, the proposed sampling criteria are:
For LQ 1.25 the sample size is 500 pieces, accept on 3 defects, reject on 4 defects. D=0.27
For LQ 5.0 the sample size is 500 pieces, accept on 18 defects, reject on 19 defects. D=2.5
For LQ 8.0 the sample size is 315 pieces, accept on 18 defects, reject on 19 defects. D=3.9
For LQ 12.5 the sample size is 200 pieces, accept on 18 defects, reject on 19 defects. D=6.3
Where D = process average percent nonconforming.

Accept means that the lot is determined to be within the LQ, reject means that the lot defect rate exceeds the LQ.

LQ means that there is a 95% probability that the actual lot quality is equal or better than the stated requirement (1.25%). The process average (D), which corresponds to an AQL value, must be significantly better than the LQ in order to have a 95% probability of acceptance.
With the Q3-1988 plan and a 500-piece SR-100 sample size, information at LQs of 2.0 and 3.15 can also be obtained. For LQ 2.0 accept criteria is not more than 5 defects and for LQ 3.15 accept criteria is not more than 10 defects.

Sample Collection

NIOSH will use the current MSHA SCSR inventory of all mine-deployed SR-100 units to randomize units manufactured after October 2003. The significance of October 2003 relates to similarity in design and construction. This is the date when the threads on the SR-100 starter-oxygen bottle were modified to the current design. The heat-exposure indicator was also added to the SR-100 in that timeframe. (Fewer than 4700 of the currently deployed SR-100s in service in mines are older than this cut-off.)

Using the MSHA inventory, and the unit serial number as a unique identifier, all in-service SR-100s newer than October 2003, will be randomized. As defined in Q3-1988, a sample size of 500 is needed to ascertain the target LQ values at the stated level of confidence (95%). Past experience collecting SCSRs from mine service indicates that more than 500 units will need to be assessed in order to obtain 500 recoverable units for testing. A recoverable unit is a unit that can be found and passes all manufacturer inspection criteria for use. There are numerous reasons that individual units may not be found. Foremost among the reasons is that a unit reported in the inventory can be damaged and removed from service between the time it is recorded and the time that it is searched for in the collection. NIOSH will tabulate and organize the randomly selected unit serial numbers by MSHA District and request each District to retrieve the SR-100s on its respective list. All units not recoverable shall be noted as such on the sample inventory list provided to the Districts, and the reason for which they are deemed to be unrecoverable shall also be made part of that record. NIOSH will then retrieve units from the MSHA Districts.

The ultimate goal of the collection is to assemble the lowest-ordered-500 randomly-listed, recoverable units available for testing. Since the logistics of sample collection and retrieval are not trivial, NIOSH proposes to accumulate these from the MSHA Districts in at least two forays to reduce the amount of over-selection needed to obtain the required sample. Units retrieved by NIOSH from the MSHA district offices will be re-inspected in the NPPTL lab by a NIOSH technician and tested in an order that ensures the goal of the collection, mentioned above, is met in full.

NIOSH also wishes to reserve the ability to evaluate the test results in the most efficient manner permitted under the QA sampling plan. For instance, if 19 defects were to be observed within the first 100 units tested, the sample collection, if it has not already been completed, and testing could be halted. In that event, the logic of the QA sample plan indicates the highest selected lot rejection criterion has been exceeded at the number of observations. We would not know what the actual prevalence is, but we would know that the QA percent defective exceeds 12.5%, and a replacement rate decision could be made on that basis. Appendix A details some critical sampling plan decision points based on the number of defects observed.

Test Plan Outline

1. Record the SR-100 unit serial number and date of manufacture.
2. Determine the minimum allowable volume of starter oxygen based on unit date of manufacturer and CSE maximum allowable leakage rate.
3. Open SR-100
    a. If unable to open put the unit aside and do not use for this evaluation.
    b. If the unit opens proceed to step 4.
4. Record serial number of oxygen starter cylinder.
5. Measure volume of starter oxygen available. The breathing bag will be removed from the unit and the unit connected to a spirometer and an electronic data recorder for the volume measurement.

6. Compare measured volume of starter oxygen available with the minimum allowable oxygen volume from step 2.
7. Test result:
    a. If the volume of starter oxygen available is equal to or greater than ($\geq$) the minimum allowable starter oxygen the unit is an accept (PASS);
    b. If the volume of starter oxygen available is less than (<) the minimum allowable starter oxygen the unit is a reject (FAIL).

Appendix A
Strategy for Sampling Decisions Based on SCSR Sampling / Test Results

The sampling strategies are based upon ultimately obtaining a sample of 500 units and applying the Q3-1988 sample size and accept/reject criteria. The terminology used is standard for QA lot acceptance, i.e., accept the tested lot as meeting the stated quality level versus fail the tested lot as not meeting the stated quality level. For our purposes, acceptance translates to a measured defect rate at or below the stated quality level, and reject means the measured defect rate exceeds the stated quality level. The selected LQ Level sampling criteria are listed below:

For LQ 1.25 the sample size is 500 pieces, accept on 3 defects, reject on 4 defects.
For LQ 5.0 the sample size is 500 pieces, accept on 18 defects, reject on 19 defects.
For LQ 8.0 the sample size is 315 pieces, accept on 18 defects, reject on 19 defects.
For LQ 12.5 the sample size is 200 pieces, accept on 18 defects, reject on 19 defects.

With the sample size goal of 500 units there are several possible sample collection decision points based on the Limiting Quality (LQ) sampling criteria, the actual in-process number of units sampled and tested, and the number of defects found. Where N = the number of units sampled and tested and X = the number of defects detected, the decision points are:

A. While $N \leq 200$ units:

    a. If $X \leq 18$, defect rate is less than 12.5%, sampling and testing continues.
    b. If $X = 19$, defect rate is at least 12.5%, test results will be evaluated and reported.

B. When $201 \leq N \leq 315$:

    a. If $X \leq 18$, defect rate is less than 8.0%, sampling and testing continues.
    b. If $X = 19$, defect rate is at least 8.0%, test results will be evaluated and reported.

C. While $316 \leq N < 500$ units:

    a. If $X \leq 18$, defect rate is less than 5.0%, sampling and testing continues.
    b. If $X = 19$, defect rate is at least 5.0%, test results will be evaluated and reported.

D. When $N = 500$ units:

    a. $X \leq 3$, defect rate is less than 1.25%.
    b. $X \geq 4$, defect rate is at least 1.25%.

The table in Attachment 1 provides a summary of the actual rate defective cut-off points and impact on redundancy.

Attachment 1

Condition	Measured Defect Rate, R	Effective Failure Rate w/redundancy
A	R <=1.25%	less than 1 in 10,000
B	R <=3.0%*	less than 1 in 1000
C	3.0%< R <5.0%	at least 1 in 1000
D	R=>5.0%	at least 2 in 1000
E	R=>8.0%	at least 6 in 1000
F	R=>12.5%	at least 1 in 100

*A measured defect rate of 3% represents the approximate point at which the effective failure rate in any sequential selection of two units rises above 1 in 1000.

www.ingramcontent.com/pod-product-compliance
Lightning Source LLC
Chambersburg PA
CBHW081824170526
45167CB00008B/3529